Selvaraj Ranganathan
Ganesh Moorthy Innasi Muthu

Green Fuel Technology

Microbial Oil Production from
Oleaginous Yeast (*Cryptococcus curvatus*)

Anchor Academic
Publishing

Ranganathan, Selvaraj, Innasi Muthu, Ganesh Moorthy: Green Fuel Technology.
Microbial Oil Production from Oleaginous Yeast (*Cryptococcus curvatus*), Hamburg,
Anchor Academic Publishing 2016

Buch-ISBN: 978-3-96067-044-5
PDF-eBook-ISBN: 978-3-96067-544-0
Druck/Herstellung: Anchor Academic Publishing, Hamburg, 2016

Bibliografische Information der Deutschen Nationalbibliothek:
Die Deutsche Nationalbibliothek verzeichnet diese Publikation in der Deutschen
Nationalbibliografie; detaillierte bibliografische Daten sind im Internet über
http://dnb.d-nb.de abrufbar.

Bibliographical Information of the German National Library:
The German National Library lists this publication in the German National Bibliography.
Detailed bibliographic data can be found at: http://dnb.d-nb.de

All rights reserved. This publication may not be reproduced, stored in a retrieval system
or transmitted, in any form or by any means, electronic, mechanical, photocopying,
recording or otherwise, without the prior permission of the publishers.

Das Werk einschließlich aller seiner Teile ist urheberrechtlich geschützt. Jede Verwertung
außerhalb der Grenzen des Urheberrechtsgesetzes ist ohne Zustimmung des Verlages
unzulässig und strafbar. Dies gilt insbesondere für Vervielfältigungen, Übersetzungen,
Mikroverfilmungen und die Einspeicherung und Bearbeitung in elektronischen Systemen.

Die Wiedergabe von Gebrauchsnamen, Handelsnamen, Warenbezeichnungen usw. in
diesem Werk berechtigt auch ohne besondere Kennzeichnung nicht zu der Annahme,
dass solche Namen im Sinne der Warenzeichen- und Markenschutz-Gesetzgebung als frei
zu betrachten wären und daher von jedermann benutzt werden dürften.

Die Informationen in diesem Werk wurden mit Sorgfalt erarbeitet. Dennoch können
Fehler nicht vollständig ausgeschlossen werden und die Diplomica Verlag GmbH, die
Autoren oder Übersetzer übernehmen keine juristische Verantwortung oder irgendeine
Haftung für evtl. verbliebene fehlerhafte Angaben und deren Folgen.

Alle Rechte vorbehalten

© Anchor Academic Publishing, Imprint der Diplomica Verlag GmbH
Hermannstal 119k, 22119 Hamburg
http://www.diplomica-verlag.de, Hamburg 2016
Printed in Germany

TABLE OF CONTENTS

1.	**INTRODUCTION**	6
1.1	Biofuel feedstocks	8
1.2	Synthetic Biology and Biodiesel Production	10
1.3	Indirect synthesis of Biodiesel from Microbial oils	10
1.4	Direct synthesis of Biodiesel from Microbial oils	11
2.	**REVIEW OF LITERATURE**	12
2.1	Medium optimization for the Production of Microbial oil	12
	2.1.1 Medium optimization	12
2.2	Raw material for the Production of microbial oil	13
	2.2.1 Yeast as the source for Production	13
	2.2.2 Folch Extraction method	15
	2.2.3 Transesterification method	16
	2.2.4 Chemistry of Transesterification reaction	17
2.3	Properties of Biodiesel	17
	2.3.1 Cetane Number	17
	2.3.2 Viscosity	18
	2.3.3 Flash point	18
	2.3.4 Pour point	19
	2.3.5 Cloud point	19
	2.3.6 Iodine number and Polyunsaturated methyl ester	20

3.	**MATERIALS AND METHODS**		20
	3.1	Microorganism	20
	3.2	Culture of yeast	22
	3.3	Staining Method	23
	3.4	Designing of Media	23
	3.5	Microbial oil production	26
	3.6	Estimation of Microbial oil	26
	3.7	Folch Extraction Method	27
	3.8	Transesterification	27
	3.9	Confirmation test for Microbial oil	28
	3.10	Confirmation test for Biodiesel	28
4.	**RESULTS AND DISCUSSION**		28
	4.1	Medium optimization	28
		4.1.1 Plackett Burmann Method	28
		4.1.2 Response Surface Method	31
	4.2	Confirmation result for Microbial oil	35
	4.3	Confirmation result for Biodiesel	35
	4.4	Discussions	36
5.	**CONCLUSION**		36
6.	**REFERENCES**		37

LIST OF ABBREVATIONS

FAEE	Fatty Acid Ethyl Ester
RSM	Response Surface Method
TAG	Triacylglycerides
STE$_s$	Steryl Esters
FAME$_s$	Fatty Acid Methyl Esters
FAAE$_s$	Fatty Acid Alkyl Esters
OR	Operations Research
CPM	Cost per thousand household exposures to this Media Vehicle
FFA	Free Fatty Acid
MTCC	Microbial Tissue Culture Collection
AOAC	Association of Official Analytical Chemists
KH$_2$PO$_4$	Potassium dihydrogen phosphate
MgSO4.7H2O	Magnesium sulphate
NaCl	Sodium Chloride
CaCO$_3$	Calcium Carbonate
KCl	Potassium chloride
CO	Carbon Monoxide
HC	HydroCarbons

ABSTRACT

The increased industrialization and motorization has led to a steep rise for the demand of petroleum products. Petroleum based fuels are obtained from limited reserves. Hence, it is necessary to look for alternative fuels, which can be produced from oleaginous yeast *Cryptococcus curvatus.* The objective of the project is medium optimization for economical production of Biomass and Microbial oil by the well known oleaginous yeast Cryptococcus *curvatus* and conversion of the lipids to biodiesel (Fatty Acid Ethyl Esters). The media was optimized using the design software STATEASE. Variables with best optimized value for the high yield of Biomass was screened by **Plackett Burmann Method**. Then the optimum concentration of those screened variables was predicted using **RSM(Response Surface Method).** The extraction method **Folch Extraction method** was used for the extraction of microbial oil .

1. INTRODUCTION

In recent times, The world has been confronted with an energy crisis due to depletion of resources and increased industrialization, motorization and increased environmental problems. on the other hand with population increasing rapidly and many developing countries expanding their industrial base and output, worldwide energy demand is bound to increase. It is estimated that the known crude oil reserves could be depleted in less than 50 years at the present rate of consumption.

Biodiesel is an efficient, non – toxic, biodegradable and clean burning fuel alternative to petroleum fuels which runs in unmodified diesel engine. Diesel fuel is largely utilized in the transport, agricultural, commercial, domestic, and industrial sectors for the generation of power / mechanical energy, and the substitution of even a small fraction of total consumption by alternative fuels will have a significant impact on the economy and the environment

Global warming and the continued depletion of non-renewable fuel resources are two major problems that entangle our planet today and demand immediate solutions . The extensive use of fossil fuels has caused greenhouse gas emissions and damage to the environment, and has also led to the current instability of oil supplies and continuous fluctuations in prices. These factors, which revolve around economic, environmental and geopolitical issues, are central to the continued interest seen in renewable energy sources . An entire branch of biotechnology, referred to as "white biotechnology" , centers on the bioproduction of fuels and chemicals from renewable sources. For biofuels, delicate optimization, and fine tuning of these processes to maximize productivity and yield is of particular concern, as the viability of any biofuel process is extremely sensitive to factors related to both raw material supply and production costs.

About 90% of the current biofuel market is represented by biodiesel and bioethanol. However, bioethanol is not seen as an ideal biofuel for the future because of its low energy density and incompatibility with the existing fuel infrastructure . On the contrary, biodiesel is already better established and is preferable to petrodiesel in terms of several characteristics, such as environmental friendliness, renewability, reduced emissions, higher combustion efficiency, improved lubricity, and higher levels of safety . Chemically, biodiesel comprises a mixture of Fatty Acid Alkyl Esters (FAAEs). The most commonly used method to produce biodiesel is the *in vitro* transesterification process, where Triacylglycerides (TAGs) of vegetable oils are combined with methanol to form Fatty Acid Methyl Esters (FAMEs) and the byproduct glycerol Alkalies (e.g., sodium hydroxide, potassium hydroxide, sodium metoxide, and potassium metoxide) , acids (e.g., sulfuric acid) , or enzymes can be used to catalyze this reaction . However, issues related to high cost and limited availability of vegetable oils have become growing concerns for large-scale commercial viability of biodiesel production . Also, the *in vitro* transesterification reaction presents some unresolved issues, such as the need to use large amounts of toxic compounds (sodium hydroxide, sulfuric acid, or methanol) and the high cost of isolation and immobilization of enzyme catalysts . Various approaches to addressing these problems have been explored. First, increasing interest in developing microbial processes for the production of biodiesel from a wide range of other raw materials may represent a promising alternative to the vegetable oils. Second, technologies now exist that use living cells to synthesize products that are more easily biodegradable, require less energy, and create less waste during production than those obtained by chemical synthesis. In order for a fermentation process to compete with existing petroleum-based processes, the target molecule must be produced at high levels of yield, titer, and productivity. These goals can be difficult to attain with naturally occurring microbes. While metabolic engineering has enabled extraordinary advances in the redesign of pathways for efficient target molecule production, including biofuels, tools from synthetic biology make it possible to create new biological functions that do not exist in nature. Essentially, this is achieved either by heterologous expression of natural pathways or design of *de novo* pathways. This paper reviews approaches to microbial synthesis of biodiesel, focusing on the role of synthetic biology as an enabling technology in the design of optimal cell factories.

1.1 Biofuel feedstocks

Because of its abundance and renewable nature, biomass has the potential to produce extensive supplies of reliable, affordable, and environmentally sound biofuels to replace fossil fuels. Many biomass feedstocks, which include lignocellulosic agricultural residues as well as edible and nonedible crops, can be used for the production of biofuels.

More than 95% of global biodiesel production now begins from virgin edible vegetable oils which account for about 80% of the total production cost. However, the socioeconomic impacts of large-scale biodiesel production from edible feedstocks can be significantly lowered by the use of alternative feedstocks such as nonedible oils or lignocellulosic biomass. The use of nonedible vegetable oils is especially significant for biodiesel production in developing countries ,because of the tremendous demand for edible oils as food. Increasing attention is also now being given to the use of microbial oils as biodiesel feedstock, which are produced by certain oleaginous microorganisms .Lignocellulosic biomass, on the other hand, is the largest known renewable source of carbohydrates. It generally consists of about 25% lignin and 75% carbohydrate polymers (cellulose and hemicellulose) .These polymers, upon complete hydrolysis, yield a mixture of hexose (glucose, galactose, and mannose) and pentose (arabinose and xylose) .Synthetic biodiesel can be produced from this renewable carbon source using the Fischer-Tropsch process. Although conversion of lignocellulose into biofuels appears simple in theory, the techniques used in this field are not fully established. The main reason for this lag is the recalcitrant nature of cellulose and the toxic nature of the products of lignin degradation. Several prokaryotes and eukaryotes that have cellulolytic properties and are tolerant to the toxic products of lignin degradation have been identified . However, the yield and productivity of biofuels synthesized in this way are not sufficient to meet current energy demands. An efficient cellulolytic organism should be able to hydrolyze lignocellulose completely, ferment all sugars of lignocellulosic hydrolysate simultaneously, and tolerate toxic compounds of lignin without compromising productivity .Therefore, for cost-effective production of biofuels, the fuel-producing hosts must be designed.

In terms of ethanol production, starches (maize, wheat, barley, etc.) and sugar-rich biomass (grasses, maize leaves, beets, sugar cane, etc.) have been the feedstocks most commonly used for their bioconversion .However, advances in metabolic engineering and synthetic biology have provided new tools for creating desirable phenotypes for the production of ethanol from lignocellulosic biomass .Since ethanol is one of the substrates used for *in vivo* synthesis of biodiesel, advances in terms of maximization of two-carbon alcohol production from the most economically viable feedstocks will be discussed later in this review.

1.2 Synthetic Biology and Biodiesel Production

Synthetic biology emerged around the year 2000 as a new biological discipline, and many different definitions have been applied to this field. However, one commonly used way to describe synthetic biology is as the design and construction of new biological functions that are not found in nature. Synthetic biology is a discipline encompassing contributions from many fields , but this review places particular emphasis on the design of microbes, either by modification of existing pathways or by heterologous expression of natural pathways, in order to allow efficient production of biodiesel. In this connection, the synthesis of biodiesel using microbes is currently a highly promising alternative to conventional technologies. Microbial biodiesel production has been approached from two different angles: (1) by indirect synthesis from microbial oils, which are produced and harvested for use in the conventional *in vitro* transesterification process, and (2) by direct biodiesel synthesis using redesigned cell factories to increase production of alcohols and/or FFAs, which are subsequently used for *in vivo* synthesis of biodiesel. In the following sections, both approaches are reviewed.

1.3 Indirect Synthesis of Biodiesel from Microbial Oils

It is well known that many microbes, including certain types of microalgae, bacteria, filamentous fungi, and yeasts, can accumulate intracellular lipids, primarily TAGs, with these representing a large proportion of their biomass . Oils derived from these oleaginous microbes represent promising raw materials for biodiesel production through transesterification using the plant-based process.

The use of microbial oils offers several advantages when compared to plant-derived oils . However, oleaginous microbes have varying prospects in the biodiesel industry. For example, microalgae are photoautotrophic microorganisms that can convert CO_2 directly to lipids, which can then be used for biofuel production, particularly for biodiesel . The oil content of microalgae usually ranges between 15 and 70% by weight of the dry biomass . Scaling-up process for autotrophic microalgae is complex, however, since light is needed during cultivation. Although it is known that algae could be grown in dedicated artificial ponds for generating biodiesel , the harvesting of miles and miles of algae growth is required in order to generate substantial amounts of biodiesel. Thus, while the microbiological aspects of this approach are extremely promising, the engineering aspects pose the greatest challenge. Tools from synthetic biology have been effectively used to convert certain autotrophic microalgae into heterotrophic microorganisms . Essentially, this consists of the introduction of nonnatural metabolic pathways into the autotrophic microalgae, thereby, allowing cultivation using an organic carbon source instead of photosynthesis from sunlight.

1.4 Direct Synthesis of Biodiesel from Microbial Oils

Methanol, conventionally used as part of *in vitro* transesterification, is largely derived from nonrenewable natural gas and is also both toxic and hazardous. On the contrary, ethanol can be naturally produced from renewable resources, while exhibiting low levels of toxicity and a higher degree of biodegradability. Ethanol produced endogenously can therefore be used for *in vivo* synthesis of Fatty Acid Ethyl Esters (FAEEs) with exogenously added FFAs. Similarly, microbial FFAs can be used as feedstock for *in vivo* production of biodiesel, instead of TAGs from vegetable oil.

2. REVIEW OF LITERATURE

2.1 Medium optimization for the Production of Microbial oil

2.1.1 Medium optimization

The idea of optimization derives from an engineering discipline called "Operations Research" and known as "OR". OR consists of a set of tools and approaches known respectively as "algorithms" and "heuristics". Algorithms are mathematical equations, while heuristics are fuzzier methods i.e. they are not equations. Both algorithms and heuristics are aimed at improvement in the operations of an organization. "Optimization models" are a type of algorithm intended to provide the best possible solution to some problem facing an organization. Where the problem itself is so complex that finding the best possible solution could cost more than the benefit of doing so, the optimization models generally do not attempt to find the best possible solution, but instead seek to find extremely good solutions within reasonable cost and time parameters. This in fact is the more common situation. Although in the latter case what is sought is literally "improvement" rather than "optimization", these models are still conventionally called optimization models in all cases. Within the guts of the model, all information about a particular media vehicle are generally reduced to a single number representing its cost: benefit value. A simplified example of such a cost:benefit value is the CPM (Cost Per Thousand household exposures to this media vehicle). A slightly less simplified example is the CPM Targets (Cost Per Thousand Target Audience exposures to this media vehicle). The model is designed to maximize value by selecting those vehicles with the lowest cost in relation to whatever parameter is to be maximized. The parameter to be maximized (e.g. Target Audience exposures) is technically known as the "objective function". The builders of media optimization models studied the way media planners and buyers conventionally selected media vehicles, and constructed their models to mirror these conventional procedures. In doing so, they sought to move from the heuristics being used by planners/buyers into the use of true algorithms instead.

2.2 Raw material for the Production of Microbial oil

2.2.1 Yeast as the source for Production

The efficiency of oil biosynthesis by yeast and its composition depends on the genetic properties of the yeast strains, cultivation conditions and the composition of culture medium. Lipids are important storage compounds in yeast. Storage lipids are usually found within special organelles knows as lipid particles or lipid bodies. In yeast, these lipid bodies accumulate during stationary phase and they can constitute up to 70% of the total lipid content of the cell.

Triacylglycerols (TAGs) and Steryl Esters (STEs) are the most important storage lipids of eukaryotes cells such like yeast cells. TAG provides an energy source on one hand and a source of fatty acids for membrane phospholipid formation on the other hand. Mobilization of STE sets sterols free, which are also required for membrane proliferation, especially of the plasma membrane. In the yeast as in other eukaryotic cells, TAG and STE form the core of the so-called lipid particles which are surrounded by a phospholipids monolayer with a small amount of proteins embedded.

Lipids are soluble in organic solvents, but sparingly soluble or insoluble in water. The existing procedures for the extraction of lipids from source material usually involves selective solvent extraction and the starting material may be subjected to drying prior to extraction. Solubility of lipids is an important criterion for their extraction of microbial oil from source material and depends heavily on the type of lipid present, and the proportion of nonpolar (principally triacylglycerols) and polar lipids (mainly phospholipids and glycolipids) in
the sample; therefore, several solvent systems might be considered, depending on the type of sample and its components. The solvents of choice are chloroform/methanol or chloroform/methanol/water, in the case of the Folch Method.

The main aim of the project is medium optimization for economical production of microbial oil by the well known oleaginous yeast *Cryptococcus curvatus* and conversion of the microbial oil to biodiesel. The oil extraction process was optimized using environmentally safe solvents. The *C. curvatus* oil gets trans-esterifies to biodiesel and thoroughly characterizes.

Cryptococcus curvatus is an extremophile found in cold-seep sites. It is oleaginous, and uses the sugars in cellulose for the growth and production of storage triglycerides. Molecular evidence that phylogenetically diverged ciliates are active in microbial mats of deep-sea cold-seep sediment. This species has been extensively studied in relation ship to lipids. It can uptake both glucose and xylose simultaneously. When grown in old oil with high levels of polymerized triglyceride, the cell wall transforms from being smooth to having hair or wart-like protuberances which are believed to assist in lipid uptake.

2.2.2 Folch Extraction method

The most popular extraction procedure for Extraction of microbial oil is that of Folch (*Folch et al., J Biol Chem 1957, 226, 497*)

1. The tissue is homogenized with chloroform/methanol (2/1) to a final volume 20 times the volume of the tissue sample (1 g in 20 ml of solvent mixture). After dispersion, the whole mixture is agitated during 15-20 min in an orbital shaker at room temperature.
2. The homogenate is either filtrated (funnel with a folded filter paper) or centrifuged to recover the liquid phase.
3. The solvent is washed with 0.2 volume (4 ml for 20 ml) of water or better 0.9% NaCl solution. After vortexing some seconds, the mixture is centrifuged at low speed (2000 rpm) to separate the two phases. Remove the upper phase by siphoning and kept it to analyze gangliosides or small organic polar molecules. If necessary (need of removing labelled molecules...) rinse the interface one or two times with methanol/water (1/1) without mixing the whole preparation.
4. After centrifugation and siphoning of the upper phase, the lower chloroform phase containing lipids is evaporated under vacuum in a rotary evaporator or under a nitrogen stream if the volume is under 2-3 ml.

2.2.3 Transesterification method

Transesterification also called alcoholysis, is the displacement of alcohol from an ester by another alcohol in a process similar to hydrolysis. (Fuge RO, Gros AT) his process has been widely used to reduce the viscosity of triglycerides. The Transesterification reaction is represented by the general equation.

$$RCOOR' + R'' \rightarrow RCOOR'' + R'OH$$

If methanol is used in the above reaction, it is termed methanolysis. The reaction of triglyceride with methanol is represented by the general equation: Triglycerides are readily trans-esterified in the presence of alkaline catalyst at atmospheric pressure and at a temperature of approximately 60 to 70°c with an excess of methanol. The mixture at the end of reaction is allowed to settle.

The lower glycerol layer is drawn off while the upper methyl ester layer is washed to remove entrained glycerol and is then processed further. The excess methanol is recovered by distillation and sent to a rectifying column for purification and recycled. The Transesterification works well when the starting oil is of high quality. However, quite often low quality oils are used as raw materials for bio-diesel preparation. In cases where the free fatty acid content of the oil is above 1%, difficulties arise due to the formation of soap which promotes emulsification during the water washing stage and at an FFA content above.

2.2.4 Chemistry of Transesterification reaction

The overall Transesterification reaction is given by three consecutive and Reversible equations as below:

$$\text{Triglycerides} + \text{ROH} \xrightleftharpoons{\text{Catalyst}} \text{Diglycerides} + \text{R'COOR}$$

$$\text{Diglyceride} + \text{ROH} \xrightleftharpoons{\text{Catalyst}} \text{Monoglyceride} + \text{R''COOR}$$

$$\text{Monoglyceride} + \text{ROH} \xrightleftharpoons{\text{Catalyst}} \text{Glycerol} + \text{R'''COOR}$$

The first step is the conversion of triglycerides to diglycerides, followed by the conversion of diglycerides to monoglycerides, and of monoglycerides to glycerol, yielding one methyl ester molecule per mole of glyceride at each step.

2.3 Properties of Biodiesel

A general understanding of the various properties of Biodiesel is essential to study implications in engine use, storage, handling and safety.

2.3.1 Cetane Number

Cetane number of a diesel engine fuel is indicative of its ignition characteristics. Higher the cetane number better it is in its ignition properties. Cetane number affects a number of engine performance parameters like combustion, Stability, drivability, White smoke, noise and emissions of CO and HC. Bio-diesel has higthe her cetane number than conventional diesel fuel. This results in higher combustion efficiency and smoother combustion. No correlation was found between Specific gravity and the cetane number of various biodiesel (M.Graboski & H.Shapouri, 1994). It is important to note that cetane index, commonly used to indicate the ignition characteristics of diesel fuels, does not give correct results for biodiesel.(K.Pramanik.,2000). Hence Cetane index is not specified and a cetane number test is necessary. Even for a biodiesel blend, cetane index is not applicable as it does give a correct approximation of cetane number of the blend.

2.3.2 Viscosity

In addition of lubrication of fuel injection system components, Fuel viscosity controls the characteristics of the injection from the diesel injector (droplet size, spray characteristics). The viscosity of methyl esters can go to very high levels and hence, it is important to control it within an acceptable level to avoid negative impact on fuel injection system performance (M.A.Kalam,H.H.Masjuki, 1998) Therefore, the viscosity specifications proposed are same as that of the diesel fuel.

2.3.3 Flash point

Flash point of a fuel is defined as the temperature at which it will ignite when exposed to a flame or spark. The flashpoint of bio-diesel is higher than the petroleum based diesel fuel. Flashpoint of bio-diesel blends is dependent on the flashpoint of the base diesel fuel used, and increase with percentage of bio-diesel in the blend . Thus, in storage , biodiesel and its blends are safer than conventional diesel. The flashpoint of bio-diesel is around 160°c, but it can reduce drastically if the alcohol used in manufacture of bio-diesel is not removed properly. Residual alcohol in the bio-diesel reduces its flashpoint drastically and is harmful to fuel pumps, seals, elastomers etc. It also reduces the combustion quality. (Booz-Allen & Hamilton, Inc)

A minimum flashpoint for biodiesel is specified more from the point of view of restricting the alcohol content rather than a fire hazard. A minimum flashpoint of 100°c is specified to ensure that excess methanol used for the esterification is removed. Another important consideration is that the test method used to find out flashpoint (ASTM D 93) gives high scatter in results at the flashpoint nears100 °C. Due to this reason, the ASTM D 6751 standard issued in Feb, 2002 calls for a flashpoint of min. 130 °c though the intent is to get a min. value of 100°c (as specified in 1999 Draft Standard PS 121)

2.3.4 Pour Point

Normally either pour point or **CFPP** are specified for normal biodiesel. French and Italian bio-diesel specifications specify pour point whereas other specify CFFP. Since CFFP reflects more accurately the cold weather operation of fuel, it is proposed not to specify pour point for bio-diesel. Pour point depressants commonly used for diesel do not work for biodiesel.

2.3.5 Cloud point

Cloud point is the temperature at which a cloud or haze of crystals appear in the fuel under test conditions and thus becomes important for low temperature operations. Biodiesel generally has higher cloud point than diesel fuel. Cloud point limit is not specified but ASTM D 6751 calls for reporting of the cloud point to alert the user of possible problem under cold climatic conditions.

2.3.6 Iodine Number and polyunsaturated methyl ester

In diesel engines, methyl esters have been known to cause engine oil dilution by the fuel. A high content of unsaturated fatty acids in the ester (indicated by high iodine number) increases the danger of polymerization in the engine oil. Oil dilution decreases oil viscosity .Sudden increase in oil viscosity, as encountered in several engine tests, is attributed to oxidation and polymerization of unsaturated fuel parts entering into oil through dilution. In saturated fatty acids all the carbon is bound to 1997 two hydrogen atoms by double bonds. More the double bonds the lower is the cloud point of oil. The tendency of the fuel to be unstable can be predicted by iodine number (Lepori, W.A. & E.R.Engler,1997).

3. MATERIALS AND METHODS

3.1 Microorganism

The yeast strain *Cryptococcus curvatus* was obtained from MTCC, Chandigarh, India. MTCC code: 2698 .The Growth condition is Aerobic. This yeast is reported able to grow upto a temperature about 25°C(Ross A, Taylor IE(1981)).Incubation time is 48 hours. The given medium composition for the yeast strain is

TABLE 1: Media composition of yeast

Mannitol	1 gm
Potassium dihydrogen phosphate	0.05 gm
Magnesium sulphate	0.02 gm
Sodium chloride	0.01 gm

Yeast extract	0.04 gm
Calcium carbonate	0.4 gm
Agar agar	1.5 gm
Distilled water	100 ml

The above mentioned media was modified for the high yield biomass. The modified media composition is given below

TABLE 2 : Modified Media composition

Mannitol	1 gm
Potassium dihydrogen phosphate	0.05 gm
Magnesium sulphate	0.02 gm
Sodium chloride	0.01 gm
Yeast extract	0.04 gm
Calcium carbonate	0.4 gm
Potassium chloride	1.5 gm
Distilled water	100 ml

3.2 Culture of yeast

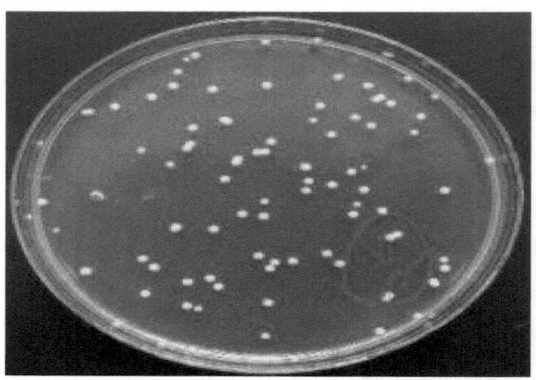

Fig 1: *Cryptococcus curvatus (Fonseca et al., 2000)*

The broth and media prepared was sterilized in autoclave.The media was poured on petri plates to produce mother culture.The sample was inoculated in broth to produce mother Culture and it was placed in shaker incubator at room temperature. After time period of 24 hours the growth of yeast *Cryptococcus curvatus(fig.1)* was observed in the broth.The yeast was sub-cultured in plates containing agar.Presence of Yeast in the was confirmed by staining method .

Fig 2: **Mother culture** (Fonseca et al., 2000)

3.3 Staining method

The smear was prepared and it was heat fixed. The smear was stained with Carbol fuchsin in the ratio of 1:3 with diluted water. After time period of 1 min, it was washed gently with tap water. Then the smear was dried using blotting paper. The smear was stained with Nigrosin (1gm in 10ml distilled water). The stain was spreaded throughout the slide using the cover slip then it was air dried. The slide was observed under microscope where the presence of an oleaginous yeast *Cryptococcus curvatus* was conformed.

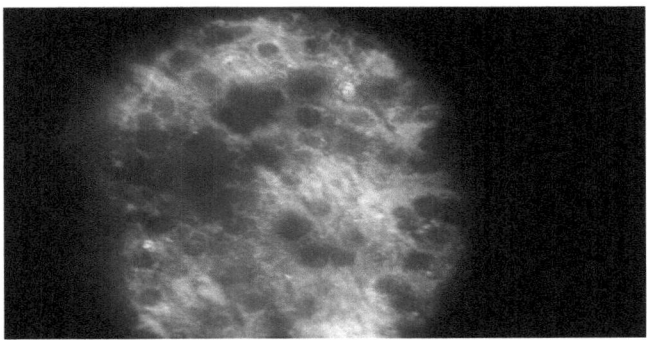

Fig 3: Microscopical view of *Cryptococcus curvatus* (Shafiee et al., 2005)

3.4 Designing of Media

The designing of media for yeast was composed using Plackett Burmann method. The plackett burmann method is one of the widely used method for medium optimization. it is done using variables of higher and lower values. The variables used for medium optimization of yeast are given as follow

TABLE 3: Media Design

COMPONENTS USED	LOWER VALUE(g/100ml)	HIGHER VALUE(g/100ml)
D-Mannitol	0.4	0.7
KH2PO4	0.03	0.04
MgSO4.7H2O	0.01	0.02
NaCl	0.008	0.01
Yeast extract	0.02	0.03
CaCO3	0.2	0.3
KCl	0.005	0.01

The above mentioned lower and higher values of media components were taken from media composition given by MTCC

TABLE 4: Plackett Burmann Design table

Trial	D-Mannitol (g/l)	KH_2PO_4 (g/l)	$MgSO_4.7H_2O$ (g/l)	NaCl (g/l)	Yeast extract (g/l)	$CaCO_3$ (g/l)	KCl (g/l)
1	0.7	0.04	0.02	0.008	0.03	0.2	0.01
2	0.4	0.04	0.02	0.01	0.02	0.3	0.005
3	0.4	0.03	0.02	0.01	0.03	0.2	0.01
4	0.7	0.03	0.01	0.01	0.03	0.3	0.005
5	0.4	0.04	0.01	0.008	0.03	0.3	0.01
6	0.7	0.03	0.02	0.008	0.02	0.3	0.01
7	0.7	0.04	0.01	0.01	0.02	0.2	0.01.
8	0.4	0.03	0.01	0.008	.02	0.2	.005

The above mentioned seven conical flask were kept in shaker incubator at room temperature for time period of 72 hours. The growth of yeast were observed in the flask according to their higher and lower values of composition. The results are to be estimated by the software STATEASE. The grown yeast were stained for morphological confirmation of *Cryptococcus curvatus*.

3.5 Microbial oil production

3 litre media was prepared by using the high yield composition(trial 2) for the production of microbial oil.

TABLE 5: High yield composition of media for the production of microbial oil

Components	Values (g/l)
Mannitol	4.00 (low)
Kh_2Po_4	0.40 (high)
$MgSo_4.7H_2O$	0.20 (high)
NaCl	0.10 (high)
Yeast Extract	0.20 (low)
$CaCo_3$	3.00 (high)
KCl	0.05 (low)

3.6 Estimation of microbial oil

Medium optimization was done using design software STATEASE version 8. yeast was cultured in 3 litre optimized media and was confirmed by the staining method , from the obtained yeast the process of extraction of microbial oil was processed. The extraction method used is FOLCH EXTRACTION METHOD.

3.7 Folch Extraction method

The sample from the 3 litre optimized culture was taken. The mixture of chloroform and methanol in the ratio of 1:2 was mixed with sample. Then the mixture was kept in shaker incubator for 20mins. The chloroform was added to the mixture and mixed for 1min. Distilled water was added and mixed for 1min. Then the mixture was centrifuged at 2000 rpm for 5 mins. The upper phase was discarded and the lower phase was collected. After evaporation, the lower phase was redissolved with chloroform and methanol in 2:1 ratio and the microbial oil was collected.

3.8 Transesterification

The above extracted microbial oil was converted into biodiesel using Transesterification process. The overall reaction of Transesterification is given as below

Methanol + NaOH catalyst + oil

⇩

R_1-cooH + CH_3OH

⇩

$RcooCH_3$

Methyl ester

3.9 Confirmation test for Microbial oil

The microbial oil was taken and it was mixed with half pellet of NaOH. Soap formation was observed which conforms the presence of microbial oil.

3.10 Confirmation test for Biodiesel

Oil was taken in cavity chamber. The cavity chamber was placed on heating mantel. The Thermometer was placed vertically on cavity chamber to determine the Flash and Fire Point. The chamber was closely watched for the appearance of the Blue flame which is the indication of the Flash point.

4. RESULTS AND DISCUSSION

4.1 Medium optimization

4.1.1 Plackett Burmann Method

By using Plackett Burmann method, we screened the optimized variables. These optimized variables were found as A-Mannitol, D-NaCl, F-CaCo3.
The plackett burmann method is one of the widely used method for medium optimization. it is done using variables of higher and lower values.

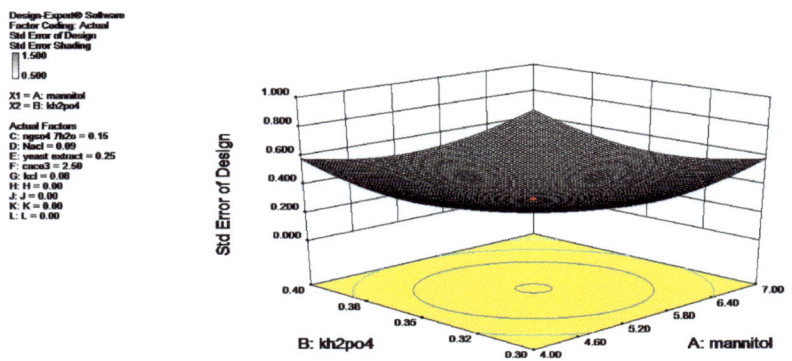

Fig. 4: 3D STRUCTURE OF PLACKETT BURMANN DESIGN OF KH$_2$PO$_4$ Vs Mannitol (Selvaraj et al., 2014)

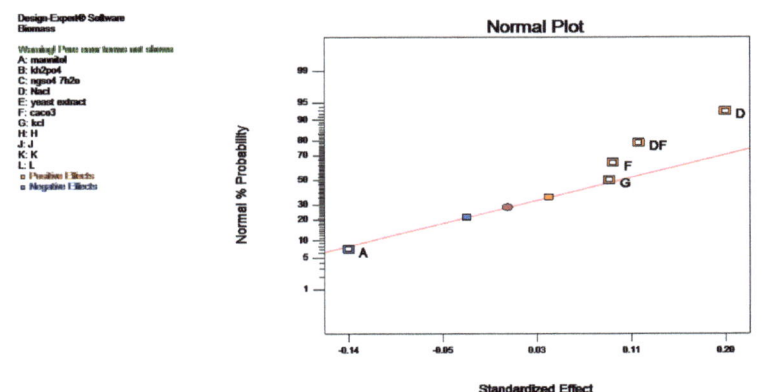

Fig. 5: Factors affecting microbial oil production

TABLE 6: ANOVA table

Sum of source	Squares	Mean df	F-value	p-value	Prob > F	
Model	0.092	5	0.018	4.52	0.0469	SIGNIFICANT
A-mannitol	0.029	1	0.029	7.19	0.0364	
D-NaCl	0.060	1	0.060	14.81	0.0085	
F-CaCo$_3$	0.015	1	0.015	3.63	0.1055	

The Model F-value of 4.52 implies the model is significant. There is only a 4.69% chance that a "Model F-Value" this large could occur due to noise. Values of "Prob > F" less than 0.0500 indicate model terms are.

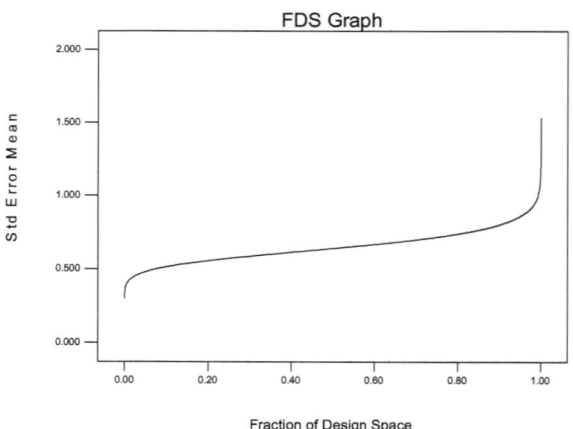

Fig. 6: FDS Graph

FDS graph gives fraction of design space. In 3D graph center points was detected which shows the values was correct and the design was possible for screening out the optimized variable. At the left side of the graph, actual design factors of the variables were predicted.

4.1.2 Response Surface Method

By using Response Surface Method, we found the optimum concentration of the above screened variables.

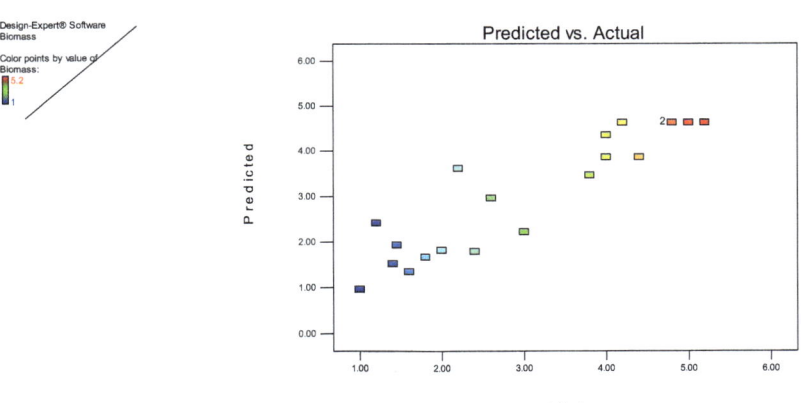

Fig. 7: Predicted Vs Actual values

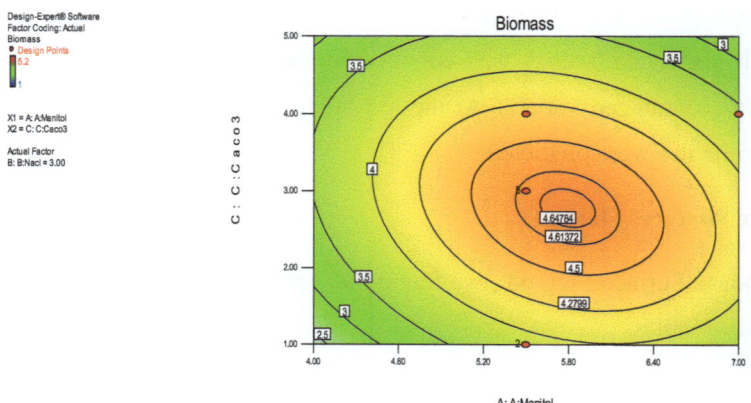

Fig. 8: Contour Plot

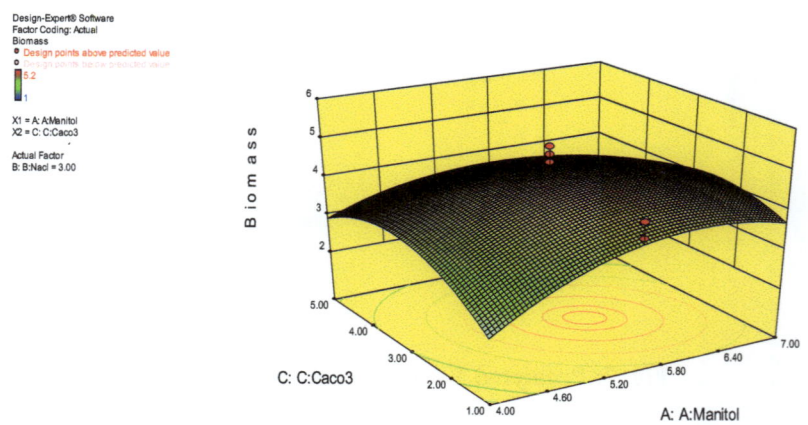

Fig. 9: 3D structure of RSM for Biomass concentration of Mannitol Vs CaCo$_3$ (Selvaraj et al., 2014)

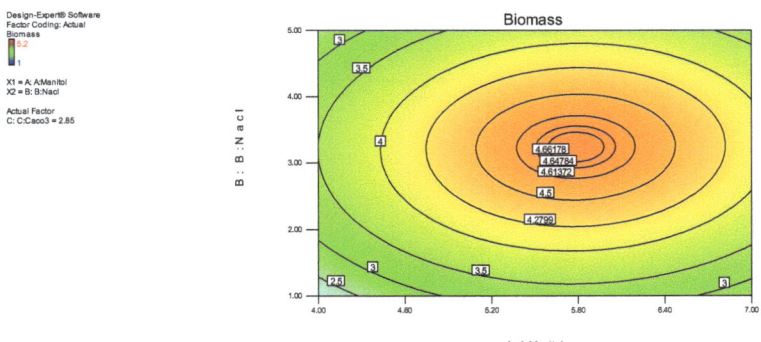

Fig. 10: Contour Plot

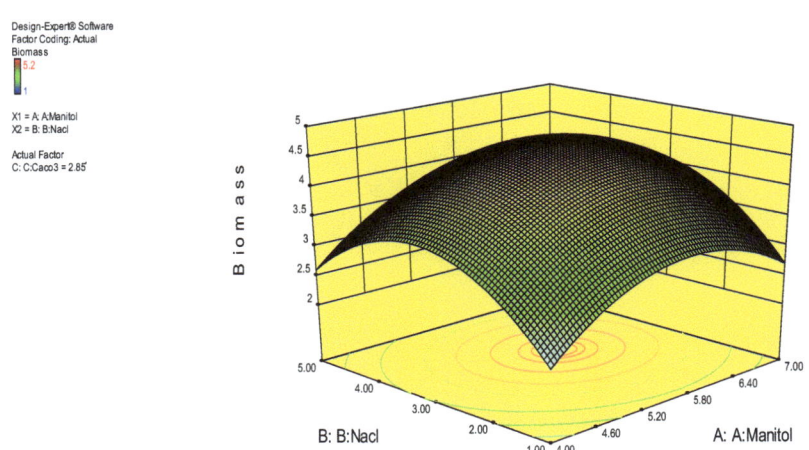

Fig. 11: Biomass concentration of Mannitol Vs NaCl (Selvaraj et al., 2014)

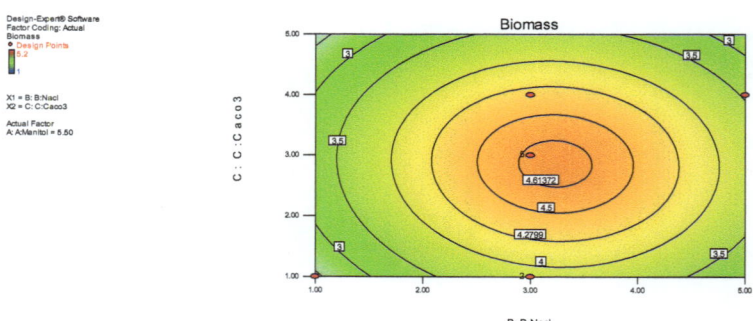

Fig. 12: Contour Plot

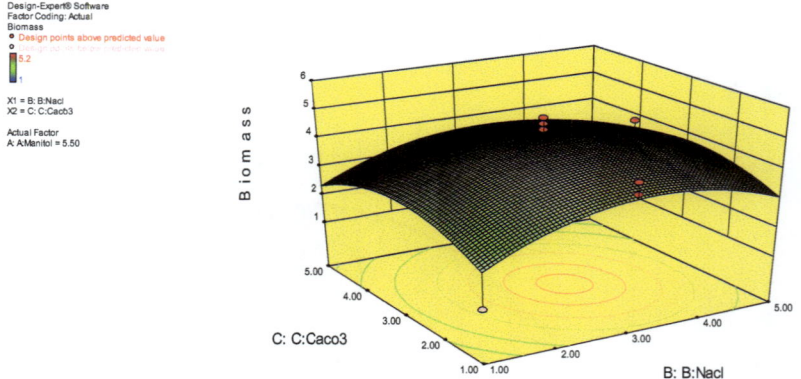

Fig. 13: Biomass concentration of NaCl Vs CaCo$_3$ (Selvaraj et al., 2014)

4.2 Confirmation result for Microbial oil

The microbial oil produced by the Folch extraction method was confirmed using the process the cetane number ,flash point,firepoint,Iodine number,pour point for

Factor	Name	Level	Low level	High level	Std Dev	coding
A	Mannitol	5.50	4.00	7.00	0.000	Actual
D	NaCl	3.00	1.00	5.00	0.000	Actual
F	CaCo3	5.00	1.00	5.00	0.000	Actual

the oil was also found

4.3 Confirmation result for Biodiesel

The microbial oil produced is converted into the biodiesel by the Transesterification process. The biodiesel produced was confirmed by their flash point and fire point.

TABLE 7: Confirmation Report

Two-sided **Confidence=95%** **n = 1**

Response	Prediction	Std Dev	SE(n=1)	95% PI low	95% PI high
Biomass	4.62766	0.786921	0.844316	2.7464	6.50891

TABLE 8: Flash point and Fire point

FLASH POINT	FIRE POINT
71°c	73°c

4.4 Discussions

As per Reference on Research journal of Microbiology Year :2009 | Volume: 4 | Issue: 8 | Page No : 301 – 313. The Biomass was 2.8g .oil production increases with the increase in Biomass. The media used here was of low cost and the media was optimized. The significant values of F test , R squared value was 0.82. Biomass achieved here was 5.5g. Thus, Biomass increases as 2 fold.

5. CONCLUSION

Insecurity in the supply of fossil fuels, volatile fuel prices, and major concerns regarding climate change have sparked renewed interest in the production of fuels from renewable resources. Because of this, the use of biodiesel has grown dramatically during the last few years and is expected to increase even further in the future. Biodiesel production through the use of microbial systems has marked a turning point in the field of biofuels since it is emerging as an attractive alternative to conventional technology. Microorganisms may therefore become an ideal platform for the production of biodiesel in the future.

References

Ambrose, M.E., Roche, B.J., and Knoble, Jr., G.M.. Semimicro method for determining total lipids in fish meal. *J. Assoc. Off. Anal. Chem* 1969.52:688-691.

AOAC (Association of Official Analytical Chemists) Official Methods of Analysis, 16th Edition. AOAC International, Gaithersburg, MD. 1995.

Atsumi S, Cann AF, Connor MR, Shen CR, Smith KM, Brynildsen MP, Chou KJ, Hanai T, Liao JC Metabolic engineering of *Escherichia coli* for 1-butanol production. Metab Eng. 2008 doi:10.1016/j.ymben.2007.08.003.

Atsumi S, Hanai T, Liao JC Non-fermentative pathways for synthesis of branched-chain higher alcohols as biofuels. Nature. 2008;451:86–89.

Baharaeen S. and Vishniac H. S. Budding Morphology of a Psychrophilic Cryptococcus and related species compared with Leucosporidium scottii. Mycologia. (1981) 73(4): 618-633

Becker DF, Fuchs JA, Banfield DK, Funk WD, MacGillivray RT, Stankovich MT. Characterization of wild-type and an active-site mutant in *Escherichia coli* of short-chain acyl-CoA dehydrogenase from *Megasphaera elsdenii*. Biochemistry. 1993;32:10736–1074

Bermejo LL, Welker NE, Papoutsakis ET. Expression of *Clostridium acetobutylicum* ATCC 824 genes in *Escherichia coli* for acetone production and acetate detoxification. Appl Environ Microbiol. 1998;64:1079–1085

Birgisson et al. Cold-adapted yeasts as producers of cold-active polygalacturonases. Extremophiles. (2003) 7:185-193

Bligh, E. G., and W. J. Dyer.. A rapid method of total lipidextraction and purification. Can. J. Biochem. Physiol. 1959 37:911-917

Boynton ZL, Bennett GN, Rudolph FB. Cloning, sequencing, and expression of genes encoding phosphotransacetylase and acetate kinase from *Clostridium acetobutylicum* ATCC 824. Appl Environ Microbiol. 1996;62:2758–2766.

Casadevall A and Perfect JR Cryptococcus neoformans. American Society for Microbiolgy, ASM Press, Washington DC, 1st edition. (1998)

Chen JS, Hiu SF. Acetone-butanol-isopropanol production by *Clostridium beijerinckii* (synonym, *Clostridium butylicum*) Biotechnol. Lett. 1986;8:371–376.

Cheng MF, Chiou CC, Liu YC, Wang HZ, Hsieh KS Cryptococcus laurentii fungemia in a premature neonate. Journal of Clinical Microbiology. (2001) 39(4):1608–11. A good review of C. laurentii cases till year 2000

Christie, W. W.. Lipid analysis, 2nd ed. Pergamon Press,
Oxford, United Kingdom. 1982

Cook. B. A note on cryptococcus after castellani. U.S. Dept. of Health. (1965)

Cryptococcus albidus. Mycology Online. The University of Adelaide.

Cryptococcus skinneri Characteristics. CBS-KNAW Fungal Biodiversity Centre. http://www.cbs.knaw.nl/collections/BioloMICS.aspx?Table=Yeasts%20species&Name=Cryptococcus%20skinneri&Fields=All&ExactMatch=T

de la Plaza M, Fernandez de Palencia P, Pelaez C, Requena T. Biochemical and molecular characterization of alpha-ketoisovalerate decarboxylase, an enzyme involved in the formation of aldehydes from amino acids by *Lactococcus lactis*. FEMS Microbiol Lett. 2004;238:367–374.

Duncombe GR, Frerman FE. Molecular and catalytic properties of the acetoacetyl-coenzyme A thiolase of *Escherichia coli*. Arch Biochem Biophys. 1976;176:159–170.

Folch, J., M. Lees, and G. H. Sloane- Stanley,A simple method for the isolation and purification of total lipids from animal tissues. J. Biol. Chem. . 1957. 226:497-509.

Fontaine L, Meynial-Salles I, Girbal L, Yang X, Croux C, Soucaille P. Molecular characterization and transcriptional analysis of adhE2, the gene encoding the NADH-dependent aldehyde/alcohol dehydrogenase responsible for butanol production in alcohologenic cultures of *Clostridium acetobutylicum* ATCC 824. J Bacteriol. 2002;184:821–830.

Fonseca A., Scorzeti G., and Fell J. (2000) Diversity in the yeast Cryptococcus albidus and related species as revealed by ribosomal DNA sequence analysis. Can. J. Microbiol. 46:7-27

Hanai T, Atsumi S, Liao JC Engineered synthetic pathway for isopropanol production in Escherichia coli. Appl Environ Microbiol. 2007;73:7814–7818.

http://www.doctorfungus.org/thefungi/Cryptococcus_albidus.php Results from a search on Doctor Fungus

http://www.pubmedcentral.nih.gov/articlerender.fcgi?tool=pmcentrez&artid=2725802
Ismaiel AA, Zhu CX, Colby GD, Chen JS. Purification and characterization of a primary-secondary alcohol dehydrogenase from two strains of *Clostridium beijerinckii*. J Bacteriol. 1993;175:5097–5105.

Jenkins LS, Nunn WD. Genetic and molecular characterization of the genes involved in short-chain fatty acid degradation in *Escherichia coli*: the ato system. J Bacteriol. 1987;169:42–52.

Jones DT, Woods DR. Acetone-butanol fermentation revisited. Microbiol Rev. 1986;50:484–524.

Kalscheuer R, Stolting T, Steinbuchel A Microdiesel: *Escherichia coli* engineered for fuel production. Microbiology. 2006;152:2529–2536.

Kisumi M, Sugiura M, Chibata I. Biosynthesis of norvaline, norleucine, and homoisoleucine in *Serratia marcescens*. J Biochem. 1976;80:333–339.

Labrecque O., Sylvestre D. and Messier S. Systemic Cryptococcus albidus infection in a Doberman Pinscher. J Vet Diagn Invest (2005) 17:598-600

Lamed RJ, Zeikus JG. Novel NADP-linked alcohol--aldehyde/ketone oxidoreductase in thermophilic ethanologenic bacteria. Biochem J. 1981;195:183–190.

Lee, C.M., Trevino, B., and Chaiyawat, M.. simple and rapid solvent extraction method fordetermining total lipids in fish tissue. *JAOCSInternational* 1996 79:487-492.

Leeuw et al The effects of palm oil breakdown products on lipid turnover and morphology of fungi. Can. J. Microbiol. . (2010) 56: 883-889

Lin YL, Blaschek HP. Butanol Production by a Butanol-Tolerant Strain of *Clostridium acetobutylicum* in Extruded Corn Broth. Appl Environ Microbiol. 1983;45:966–973.

Lindberg J, Hagen F, Laursen A, et al. "Cryptococcus gattii Risk for Tourists Visiting Vancouver Island, Canada". Emerg Infect Dis 13 (1): . (2007) 178–179.

MacDougall L, Kidd SE, Galanis E, et al. "Spread of Cryptococcus gattii in British Columbia, Canada, and Detection in the Pacific Northwest, USA". Emerg Infect Dis 13 (1): . (2007) 42–50.

Menna M. (Cryptococcus terreus n.sp., from Soil in New Zealand. J. gen. Microbiol.) (1954) 11:195-197

Miwa K, Tsuchida T, Kurahashi O, Nakamori S, Sano K, Momose H. Construction of L-Threonine Overproducing Strains of *Escherichia coli* K-12 Using Recombinant. Agric. Biol. Chem. 1983;47:2329–2334.

O'Neill H, Mayhew SG, Butler G. Cloning and analysis of the genes for a novel electron-transferring flavoprotein from *Megasphaera elsdenii*. Expression and characterization of the recombinant protein. J Biol Chem. 1998;273:21015–21024.

Osburn OL, Brown RW, Werkman CH. The butyl alcohol-isopropyl alcohol fermentation. J. Biol. Chem. 1937;121:685–695.

Peretz M, Bogin O, Tel-Or S, Cohen A, Li G, Chen JS, Burstein Y. Molecular cloning, nucleotide sequencing, and expression of genes encoding alcohol dehydrogenases from the thermophile *Thermoanaerobacter brockii* and the mesophile *Clostridium beijerinckii*. Anaerobe. 1997;3:259–270.

Pfeiffer et al., Mycocin production in cryptococcus aquaticus. Antonie van Leeuwenhoek. (2004) 86:369-375

purzetti et al.,(2000) Cryptococcus adeliensis sp. nov., a xylanase producing basidiomycetous yeast from Antarctica. Antonie van Leeuwenhoek. 77:153-157

Ross A, Taylor IE Extracellular glycoprotein from virulent and avirulent Cryptococcus species. Infection and Immunity. (1981) 31(3):911–8

R. Selvaraj, B. Bharathiraja, S. Palani . Enhanced Production Of Bacteriocin From Probiotics Using Optimization Techniques By Response Surface Methodology.10.17660/ActaHortic.2014.1054.31.

Sacchettini JC, Poulter CD. Creating isoprenoid diversity. Science. 1997;277:1788–1789.

Sentheshanuganathan S. The mechanism of the formation of higher alcohols from amino acids by *Saccharomyces cerevisiae*. Biochem J. 1960;74:568–576.

Scorzetti et al., Cryptococcus adeliensis sp. nov., a xylanase producing basidiomycetous yeast from Antarctica. Antonie van Leeuwenhoek. (2000) 77:153-157

Shafiee R., Nahvi I. and Emtiazi G. Bioconversion of Raw Starch to SCP by Coculture of Cryptococcus aerius and Saccharomyces cerevisiae. Journal of Biological Sciences. (2005) 5(6):717-723

Vaneechoutte M, Young DM, Ornston LN, De Baere T, Nemec A, Van Der Reijden T, Carr E, Tjernberg I, Dijkshoorn L. Naturally transformable *Acinetobacter* sp. strain ADP1 belongs to the newly described species *Acinetobacter baylyi*. Appl Environ Microbiol. 2006;72:932–936.

Vishniac H. S. Cryptococcus socialis sp. nov. and Cryptococcus consotionis sp. nov., Antarctic Basidoblastomycetes. Int. J. Syst. Bacteriol. (1985) 35(1):119-122

Vishniac H. S., Kurtzman C. P. Cryptococcus anarcticus sp. nov. and Cryptococcus albidosimilis sp. nov., Basidioblasomycetes from Antarctic Soils. Int. J. Syst. Bacteriol. (1992) 42(4) 547-553

Vishniac and Hempfling. Cryptooccus vishniacii sp. nov., an Antarctic Yeast. Int. J. Syst. Bacteriol. (1979) 29(2):153-158

Wallace KK, Bao ZY, Dai H, Digate R, Schuler G, Speedie MK, Reynolds KA. Purification of crotonyl-CoA reductase from *Streptomyces collinus* and cloning, sequencing and expression of the corresponding gene in *Escherichia coli*. Eur J Biochem. 1995;233:954–962.

Waddington, D. Applications of wide-line NMR, p. 1986. 341- 400.

Wiesenborn DP, Rudolph FB, Papoutsakis ET. Thiolase from *Clostridium acetobutylicum* ATCC 824 and Its Role in the Synthesis of Acids and Solvents. Appl Environ Microbiol. 1988;54:2717–2722.

Withers ST, Gottlieb SS, Lieu B, Newman JD, Keasling JD Identification of isopentenol biosynthetic genes from Bacillus subtilis by a screening method based on isoprenoid precursor toxicity. Appl Environ Microbiol. 2007;73:6277–6283.